DARWINISM

TODAY

A
DARWINIAN
LEFT

....

POLITICS, EVOLUTION AND
COOPERATION

Peter Singer

Yale University Press
New Haven and London

Published 1999 in Great Britain by Weidenfeld & Nicolson.
Published in the United States 2000 by Yale University Press.
Copyright © 1999 by Peter Singer.
Foreword copyright © 1999 by Helena Cronin and Oliver Curry.
All rights reserved.

Printed in the United States of America.

Library of Congress Cataloging-in-Publication Data

Singer, Peter, 1946–
    A Darwinian Left : politics, evolution and cooperation / Peter Singer.
            p.   cm. — (Darwinism today)
    Originally published : London : Weidenfeld & Nicolson, 1999.
    Includes bibliographical references.
    ISBN 0-300-08323-8 (cloth : alk. paper)
    1. Social Darwinism. 2. Right and left (Political science). 3. Socialism. 4. Human evolution. I. Title. II. Darwinism today (New Haven, Conn.)
    HM631 .S56   2000
    320.53'1—dc21   99–057179
                                            CIP

A catalogue record for this book is available from the British Library.

The paper in this book meets the guidelines for permanence and durability of the Committee on Production Guidelines for Book Longevity of the Council on Library Resources.

10   9   8   7   6   5   4   3   2   1

The Series Editors thank
Peter Tallack for his help.

CONTENTS

....

Foreword   ix

Introduction   3
The need for a new foundation   3
What is essential to the left?   7

1 Politics and Darwinism   10
The right-wing takeover   10
Facts and values   12
How the left got Darwin wrong   19
The dream of perfectibility   24
Old tunes keep coming back   28

2 Can the Left Accept a Darwinian View of Human
Nature?   31
Unpopular ideas   31
What is fixed and what is variable in human
nature?   35
How can reformers learn from Darwin?   40

3 Competition or Cooperation?   44

Building a more cooperative society   44

The prisoner's dilemma   47

Learning from Tit for Tat   51

4 From Cooperation to Altruism?   54

The puzzle of the evolution of altruism   54

Status for what?   58

5 A Darwinian Left for Today and Beyond   60

Notes and References

FOREWORD

. . . .

Darwinism Today is a series of short books by leading figures in the field of evolutionary theory. Each title is an authoritative pocket introduction to the Darwinian ideas that are setting today's intellectual agenda.

The series developed out of the Darwin@LSE programme at the London School of Economics, where the Darwin Seminars provide a platform for distinguished evolutionists to present the latest Darwinian thinking and to explore its application to humans. The programme is having an enormous impact, both in helping to popularize evolutionary theory and in fostering cross-disciplinary approaches to shared problems.

With the publication of **Darwinism Today** we hope that the best of the new Darwinian ideas will reach an even wider audience.

Helena Cronin and Oliver Curry

DARWIN@ [LSE]

A
DARWINIAN
LEFT

· ·

POLITICS, EVOLUTION AND
COOPERATION

INTRODUCTION

The need for a new foundation

In 1874 Karl Marx read *Statism and Anarchy* by Mikhail Bakunin, one of the founders of anarchism as an international revolutionary movement and Marx's main rival for control of the First International. Marx was not passive about anything he did and, in reading Bakunin, he copied out key passages from the book, and then added his own comments. The resulting manuscript, which Marx never intended for publication, reads like a debate between the foremost anarchist and the foremost communist of the nineteenth century. Here is a section of it:

Bakunin: Universal suffrage by the whole people of representatives and rulers of the state – this is the last word of the Marxists as well as of the democratic school. They are lies behind which lurks the despotism of a governing minority, lies all the more dangerous in that this minority appears as the expression of the so-called people's will.

Marx: Under collective property, the so-called will of the people disappears in order to make way for the real will of the cooperative.

Bakunin: Result: rule of the great majority of the people by a privileged minority. But, the Marxists say, this minority will consist of workers. Yes, indeed, but of ex-workers who, once they become only representatives or rulers of the people, cease to be workers.

Marx: No more than a manufacturer today ceases to be a capitalist when he becomes a member of the municipal council.

Bakunin: And from the heights of the state they begin to look down upon the whole common world of the workers. From that time on they represent not the people but themselves and their own claims to govern the people. Those who can doubt this know nothing at all about human nature.

Marx: If Mr Bakunin were familiar just with the position of a manager in a workers' cooperative, he could send all his nightmares about authority to the devil.

The most tragic irony of the history of the past century is that the record of governments that have claimed to be Marxist shows that Marx got it wrong, and Bakunin's 'nightmares about authority' were grimly prophetic. Bakunin's own solution to the problem of authority would no doubt also have gone awry; but when he suggests that someone who holds views like those of Marx and his followers 'know nothing at all about human nature', it is hard to disagree. Nor was Marx's mistake about human nature a minor aberration. Thirty years earlier, in one of his celebrated 'Theses on Feuerbach' (VI), Marx had written:

... the human essence is no abstraction inherent in each single individual. In its reality it is the ensemble of the social relations.

It follows from this belief that if you can totally change the 'ensemble of the social relations', you can totally change human nature. This claim goes to the heart of Marxism and of more broadly marxist (with a small 'm') thinking. As a result, it affects much of the thought of the entire left.

The left needs a new paradigm. The collapse of communism and the abandonment by democratic socialist parties of the traditional socialist objective of national ownership of the means of production have deprived the left of the goals it cherished over the two centuries in which it formed and grew to a position of great political power and intellectual influence. But that is not the only reason why the left needs a new paradigm. The trade union movement has been the powerhouse and the treasury of the left in many countries. What capitalists failed to accomplish by a century of repressive measures against trade union leaders, the World Trade Organization, enthusiastically endorsed by social democrat governments around the world, is doing for them. When barriers to imports are removed, nationally based trade unions are undermined. Now when workers in high-wage countries demand better conditions, the bosses can threaten to close the factory and import the goods from China, or some other country where wages are low and trade unionists will not cause trouble. The only way for unions to maintain their clout would be for them to organise internationally; but when the discrepancies between the living standards of workers are as great as they are today between, say, Europe and China, the common interests for doing so are lacking. No one likes to see their living standards drop, but the interests of a German worker in keeping up the payments on a new car are not likely to elicit much sympathy from Chinese workers hoping to be able to afford adequate health care and education for their children.

I have no answers to the weakening of the trade union movement, nor to the problem that this poses for political parties that have derived much of their strength from that movement. My focus here is not so much with the left as a politically organised force, as with the left as a broad body of thought, a spectrum of ideas about achieving a better society. The left, in that sense, is urgently in need of new ideas and new approaches. I want to suggest that one source of new ideas that could revitalise the left is an approach to human social, political and economic behaviour based firmly on a modern understanding of human nature. It is time for the left to take seriously the fact that we are evolved animals, and that we bear the evidence of our inheritance, not only in our anatomy and our DNA, but in our behaviour too. In other words it is time to develop a Darwinian left.

What is essential to the left?

Can the left swap Marx for Darwin, and still remain left? To answer that question we have to ask another: what is essential to the left? Let me answer this question in a personal way. During the past year I have completed both a television documentary and a book about Henry Spira. That name will mean nothing to most people, but Spira is the most remarkable person I have had the privilege of working with. When he was twelve years old, his family was living in Panama. His father ran a small clothing store, which was not doing well, and to save money the family accepted an offer from a wealthy friend to stay in some rooms in his house. The house was a mansion that took up an entire city block. One day two men who worked for the owner asked Henry if he wanted to come with them when they collected rents. He went with them, and saw at first hand how the luxurious existence of his father's benefactor was financed. They went into the slums, where poor people were menaced by the armed rent collectors. At the time, Henry had no concept of 'the left', but from that day on he was part of it. Later Spira moved to the United States, where he became a Trotskyist, worked as a merchant seaman, was blacklisted during the McCarthy era and then, when he won the right to work on ships again, was a central figure in a reform group that challenged the corrupt bosses of the National Maritime Union. In 1956 he went to the South to support blacks who were boycotting their local buses because they wanted the right to sit in the same seats as white people. When Fidel Castro overthrew the Batista dictatorship, Spira went to Cuba to see the process of land reform at first hand and, on his return to America, tried to rally public support against the CIA's attempts to overthrow Castro. He left the Trotskyists because they had lost touch with reality, and he taught ghetto kids in New York's public school system. As if that wasn't enough for one lifetime, in 1973 he read an essay of mine called 'Animal Liberation' and decided that here was another group of exploited beings that needed his help. He has subsequently become, over the last twenty years, the single most effective activist of the American animal rights movement.

Spira has a knack for putting things plainly. When I asked him why he has spent more than half a century working for the causes I have mentioned, he said simply that he is on the side of the weak, not the powerful; of the oppressed, not the oppressor; of the ridden, not the rider. And he talks of the vast quantity of pain and suffering that exists in our universe, and of his desire to do something to reduce it. That, I think, is what the left is all about. There are many ways of being on the left, and Spira's is only one of them, but what motivates him is essential to any genuine left. If we shrug our shoulders at the avoidable suffering of the weak and the poor, of those who are getting exploited and ripped off, or who simply do not have enough to sustain life at a decent level, we are not of the left. If we say that that is just the way the world is, and always will be, and there is nothing we can do about it, we are not part of the left. The left wants to do something about this situation.

At this point I could go into a long discussion about the philosophical basis of the more egalitarian society to which the left should aspire. But enough books have been published on that topic to fill a medium-sized public library, and I do not want to add to that literature now. It is enough to say there are many different ideas of equality that are compatible with the broad picture of the left that I am outlining here. My own ethical position is utilitarian, and the imperative of reducing suffering flows directly from that position. Although as a utilitarian I do not value equality for its own sake, I am very conscious of the principle of diminishing marginal utility, which tells us that while a given sum of money, say £100, makes very little difference to the utility of someone who already has a lot, it may make a huge difference to the utility of someone who has very little. In a world in which the 400 richest people have a combined net worth greater than the bottom 45 per cent of the world's population – about 2.3 billion – and over a billion people live on less than US$1 per day, that principle provides enough grounds for urging us to work towards a more equal distribution of resources.

Now that I have roughly outlined what I mean by 'the left', we can turn to the politics associated with Darwinism. I shall begin by asking: where has the left traditionally stood in regard to Darwinian thinking, and why?

CHAPTER I

....

Politics and Darwinism

The right-wing takeover

Two months after the publication of *The Origin of Species*, Darwin wrote to Charles Lyell:

> I have received in a Manchester newspaper rather a good squib, showing that I have proved might is right and therefore that Napoleon is right, and every cheating tradesman is also right.

The writer of that review may have been the first to suggest that Darwin's theory could be used as an ethical justification of the right of the strong to trample over the weak. He was certainly not the last. Darwin himself rejected the idea that any ethical implications could be drawn from his work. Nevertheless, evolution became a high fashion item among late nineteenth- and early twentieth-century American capitalists. These ideas did not all come from Darwin. Herbert Spencer, who was more than willing to draw ethical implications from evolution, provided the defenders of laissez-faire capitalism with intellectual foundations that they used to oppose state interference with market forces. Andrew Carnegie acknowledged that competition 'may be sometimes hard for the individual', but justified it on the grounds that 'it is best for the race, because it insures the survival of the fittest in every department'. John D. Rockefeller Jr wrote:

> The growth of a large business is merely a survival of the fittest . . . The American Beauty rose can be produced in the splendor and fragrance which bring cheer to its beholder only by sacrificing the early buds which grow around it. This is not an evil tendency in business. It is merely the working out of a law of nature and a law of God.

The capitalists were battling in the US Supreme Court to use the Fourteenth Amendment (which forbids any state to deprive any person of life, liberty or property, without due process of law) to strike down government attempts to regulate industry. So often did the opponents of regulation appeal to Spencer, that Mr Justice Holmes felt compelled, in one judgment, to point out that 'the Fourteenth Amendment does not enact Mr Herbert Spencer's *Social Statics*.'

Facts and values

Some versions of Social Darwinism commit the fallacy of deducing values from facts. Because the theory of evolution is a scientific theory, and the gap between facts and values remains as unbridgeable as it was when David Hume first drew attention to it in 1739, we cannot conclude that the direction of evolution is 'good'. Evolution carries no moral loading, it just happens. We are no more justified in helping it on its way than we are in doing our best to slow it down or change its direction. Nor can we, as the sociobiologist E. O. Wilson once claimed, use our knowledge of evolution to discover 'ethical premises inherent in man's biological nature' or deduce universal human rights from the fact that we are mammals. In the centuries to come we may make many discoveries about human nature. We may learn what makes human beings happy or sad, what leads them to develop their capacities for knowledge, wisdom, concern for others and a harmonious existence with their fellow creatures; but ethical premises will not be among these discoveries. Even an evolved disposition – like the disposition to repay favours one has received – cannot serve as the premise of an argument that tells us, without further ethical input, what we ought to do. There may be other evolved dispositions that we ought to reject – for example, a disposition to join in group acts of violence against people who are not members of our own group. Einstein was right when he said: 'As long as we remain with the realm of science proper, we never meet with a sentence of the type "Thou shalt not lie" . . . Scientific statements of facts and relations . . . cannot produce ethical directives.'

The fact–value distinction could provide a short way of answering the question: can there be a Darwinian left? The answer would then be: since to be of the left is to hold certain values, Darwin's theory has nothing to do with whether one is left or right. So there can be a Darwinian left as easily as there can be a Darwinian right.

But that is not the end of the discussion, because not all those who have appealed to Darwin's ideas to defend a political view have attempted to deduce values from facts. Some of them have instead used the theory of evolution to argue that a particular course of action will have the best consequences, where by 'best consequences' they appeal to some widely shared values, such as greater happiness and prosperity for all, or surpassing the greatest achievements of previous civilizations. It is not too difficult to interpret the two quotations given above, from Carnegie and Rockefeller, in this manner. Carnegie can be read as suggesting that competition will make most people better off, in the long run, and Rockefeller's example of the sacrifice of small buds to produce a more beautiful rose may be an appeal to an ethic that sees overriding value in scaling the greatest possible heights of human achievement. E. O. Wilson, too, mixes his reference to ethical premises inherent in our nature with the suggestion that, in his influential book *A Theory of Justice*, the American political philosopher John Rawls has not considered the 'ultimate ecological or genetic consequences of the rigorous prosecution of its conclusions'. It is not clear what Wilson means by this, but since Rawls advocates allowing inequalities only to the extent that they benefit the worst-off group in society, it at least looks as if Wilson is suggesting that we need to consider the genetic implications of assisting the worst-off to survive. Elsewhere Wilson also argues that understanding the biological differences between men and women will make us more aware of the price we have to pay for greater equality between the sexes.

As these examples show, there are many different ways in which Darwinian thinking can be invoked in political debate, and some are more defensible than others. We have already seen the following three:

• The idea that the direction of evolution itself is 'good' or 'right'.

This can be rejected without further discussion.

• The view that social policies may, by helping the 'less fit' to survive, have deleterious genetic consequences.

This is, to put it charitably, highly speculative. The factual basis for such a claim is strongest in regard to the provision of life-saving medical treatment to people with genetically linked diseases that without treatment would kill their victims before they could reproduce. There are, no doubt, many more people with early-onset diabetes being born because of the discovery of insulin and at least some of those would not have been born if there were no public health service to provide insulin at less than its market cost. But no one would seriously propose withholding insulin from children with diabetes in order to avoid the genetic consequences of providing insulin.

Moreover, there is a huge gap between these cases of specific medical treatments for genetically influenced diseases, and the vague suggestions sometimes heard from the political right that providing financial support to people who are unemployed will make it possible for them to have children, and therefore lead to a greater representation of 'deleterious' genes in the population. Even if there were a genetic component to something as nebulous as unemployment, to say that these genes were 'deleterious' would involve value judgments that go way beyond what the science alone can tell us.

• The assertion that an understanding of human nature in the light of evolutionary theory can help us to identify the means by which we may achieve some of our social and political goals, including various ideas of equality, as well as assessing the possible costs and benefits of doing so.

This needs to be taken seriously. It does not suggest that any social policy is wrong because it is contrary to Darwinian thinking. Instead it leaves the ethical decision up to us, merely offering to provide information relevant to that decision. While some absolutist moral theories tell us that justice must be done even if the heavens fall, consequentialists like myself will always welcome information about the likely outcome of what we are proposing to do. The usefulness of the information will, of course, vary in proportion to its reliability.

Here I shall also add a fourth way in which Darwinian thinking may be relevant to a political issue:

• The debunking or discrediting of politically influential, non-Darwinian, beliefs and ideas.

All pre-Darwinian political beliefs and ideas need to be examined to see if they contain factual elements that are incompatible with Darwinian thinking. For example, Sir Robert Filmer's doctrine of the divine right of kings depended on the idea that Adam was given authority over his children, and that this authority was passed down through the eldest line of his descendants, until, in seventeenth-century England, it reached the House of Stuart. Since the theory of evolution suggests that there never was an original Adam, nor a Garden of Eden, Darwin has given us a basis for rejecting this view.

This may seem superfluous because, as John Locke pointed out three hundred years ago, there are several other grounds for dismissing Filmer's theory. But consider a different, but related view: that God gave Adam dominion over 'the fish of the sea, and the fowl of the air, and everything that moved upon the Earth'. That belief still seems to exercise some influence on our attitudes to nonhuman animals, though it is as thoroughly refuted by the theory of evolution as is the doctrine of the divine right of kings. Even more sophisticated views of the differences between humans and animals are challenged by Darwinian thinking. Both in *The Descent of Man* and in *The Expression of the Emotions in Man and Animals*, Darwin showed in great detail that there is a continuum between humans and animals, not only with respect to anatomy and physiology, but also with regard to their mental lives. Animals, he showed, are capable of love, memory, curiosity, reason and sympathy for each other. By knocking out the intellectual foundations of the idea that we are a separate creation from the animals, and utterly different in kind, Darwinian thinking provided the basis for a revolution in our attitudes to nonhuman animals. Sadly that revolution did not occur, and despite some recent progress, it still has not occurred. Darwinian political thinkers should be more inclined to recognise, and base policies on, the similarities we identify between humans and nonhuman animals.

If Darwinian thinking tells us that we have been too ready to assume a fundamental difference in kind between human beings and nonhuman animals, it could also tell us that we are too ready to assume that all human beings are the same in all important respects. While Darwinian thought has no impact on the priority we give to equality as a moral or political ideal, it gives us grounds for believing that since men and women play different roles in reproduction, they may also differ in their inclinations or temperaments, in ways that best promote the reproductive prospects of each sex. Since women are limited in the number of children they can have, they are likely to be selective in their choice of mate. Men, on the other hand, are limited in the number of children they can have only by the number of women they can have sex with. If achieving high status increases access to women, then we can expect men to have a stronger drive for status than women. This means that we cannot use the fact that there is a disproportionately large number of men in high status positions in business or politics as a reason for concluding that there has been discrimination against women. For example, the fact that there are fewer women chief executives of major corporations than men may be due to men being more willing to subordinate their personal lives and other interests to their career goals, and bio-logical differences between men and women may be a factor in that greater readiness to sacrifice everything for the sake of getting to the top.

The diverse ways in which Darwinian thinking can connect with ethics and politics mean that drawing a distinction between facts and values does not settle all the issues about the nature of a Darwinian left. While the core of the left is a set of values, there is also a penumbra of factual beliefs that have typically been associated with the left. We need to ask whether these factual beliefs are at odds with Darwinian thinking and, if they are, what the left would be like without them.

How the left got Darwin wrong

The left's understandable but unfortunate mistake in regard to Darwinian thinking has been to accept the assumptions of the right, starting with the idea that the Darwinian struggle for existence corresponds to the vision of nature suggested by Tennyson's memorable (and pre-Darwinian) phrase, 'nature red in tooth and claw'. From this position it seemed only too clear that, if Darwinism applies to human social behaviour, then a competitive marketplace is somehow justified, or shown to be 'natural', or inevitable.

We cannot blame the left for seeing the Darwinian struggle for existence in these ruthless terms. Until the 1960s evolutionary theorists themselves neglected the role that cooperation can play in improving an organism's prospects of survival and reproductive success. John Maynard Smith has said that it was 'largely ignored' until the 1960s. So the fact that nineteenth-century Darwinism was more congenial to the right than the left is due, at least in part, to the limitations of Darwinian thinking in that period.

There was one great exception to the statement that the left accepted the 'nature red in tooth and claw' view of the struggle for existence. The geographer, naturalist and anarcho-communist Peter Kropotkin argued in his book *Mutual Aid* that Darwinists (though not always Darwin himself) had overlooked cooperation between animals of the same species as a factor in evolution. Kropotkin thus anticipated this aspect of modern Darwinism. Nevertheless, he went astray in trying to explain exactly how mutual aid could work in evolution, since he did not see clearly that for a Darwinian there is a problem in assuming that individuals behave altruistically for the sake of a larger group. Worse, for fifty years after Kropotkin wrote *Mutual Aid*, many highly respected evolutionary theorists made the same mistake. Kropotkin drew on his study of the importance of cooperation in animals and humans to argue that human beings are naturally cooperative. The crime and violence we see in human societies, he argued, are the result of governments that entrench inequality. Human beings do not need governments and would cooperate more successfully without them. Though Kropotkin was widely read, his anarchist conclusions separated him from the mainstream left, including, of course, the Marxists.

Beginning with Marx himself, Marxists have generally been enthusiastic about Darwin's account of the origin of species, as long as its implications for human beings are confined to anatomy and physiology. Since the alternative to the theory of evolution was the Christian account of divine creation, Darwin's bold hypothesis was seized on as a means of breaking the hold of 'the opium of the masses'. In 1862 Marx wrote to the German socialist Ferdinand Lassalle that:

> Darwin's book is very important and serves me as a natural-scientific basis for the class struggle in history. One has to put up with the crude English method of development, of course. Despite all

deficiencies, not only is the death-blow dealt here for the first time to 'teleology' in the sciences, but its rational meaning is empirically explained . . .

Yet Marx, consistent with his materialist theory of history, also thought that Darwin's work was itself the product of a bourgeois society:

> It is remarkable how Darwin recognises among beasts and plants his English society with its division of labour, competition, opening-up of new markets, 'inventions', and the Malthusian 'struggle for existence'.

Friedrich Engels was particularly enthusiastic about Darwin. In his speech at Marx's graveside, Engels paid Darwin the supreme compliment of comparing Marx's discovery of the law of human development with Darwin's discovery of 'the law of development of organic nature'. He even wrote a posthumously published essay entitled 'The Part Played by Labour in the Transition from Ape to Man', which attempts to blend Darwin and Marx. The essay reveals, however, that for all his enthusiasm Engels had not understood Darwin properly: since he believed that acquired characteristics could be inherited by future generations, his mode of evolution is Lamarckian rather than Darwinian. Decades later, Engels' naive support for the inheritance of acquired characteristics had tragic consequences when it was used by Soviet Lamarckians to show that their stand was consistent with Marxism and dialectical materialism.

This prepared the ground for the rise to favour, with Stalin's backing, of the pseudo-scientist T. D. Lysenko, who claimed to have made Soviet agriculture more productive by the use of Lamarckian ideas, and the dismissal, imprisonment and death of many of the leading geneticists of the Soviet Union. Under Lysenko's influence Soviet agronomy also went down the Lamarckian cul-de-sac, which certainly did not help the parlous state of Soviet agriculture.

Serious as Engels' Lamarckian lapse was, it is less fundamental a flaw than his idea that what Darwin did for natural history, Marx did for human history. In that neat characterisation there lurks the notion that Darwinian evolution stops at the dawn of human history, and the materialist forces of history take over. That idea needs to be examined more closely.

Here is Marx's own classic statement of his materialist theory of history:

The mode of production of material life conditions the social, political and intellectual life process in general. It is not the consciousness of men that determines their being, but, on the contrary, their social being that determines their consciousness.

By 'mode of production of material life' Marx meant the way in which people produce the goods that satisfy their needs – by hunting and gathering, by growing crops, by harnessing steam power to drive machines. The mode of production, he argued, gives rise to a particular set of economic relationships, such as lord and serf, or capitalist and labourer, and this economic basis determines the legal and political superstructures of society, which form our consciousness.

The materialist theory of history implies that there is no fixed human nature. It changes with every change in the mode of production. Human nature has already changed in the past – between primitive communism and feudalism, for example, or between feudalism and capitalism – and it can change again in the future. In less precise form this idea goes back long before Marx. In his *Discourse on the Origin of Inequality*, Rousseau dramatically presented the idea that the inauguration of private property changed everything:

> The first man who, having enclosed a piece of ground, bethought himself of saying 'This is mine,' and found people simple enough to believe him, was the real founder of civil society. From how many crimes, wars, and murders, from how many horrors and misfortunes might not any one have saved mankind, by pulling up the stakes, or filling up the ditch, and crying to his fellows: 'Beware of listening to this imposter; you are undone if you once forget that the fruits of the earth belong to us all, and the earth itself to nobody.'

To anyone who sees a continuity between human beings and our nonhuman ancestors, it seems implausible that Darwinism gives us the laws of evolution for natural history but stops at the dawn of human history. In his *Dialectics of Nature*, Engels wrote:

23

The most that the animal can achieve is to collect; man produces, he prepares the means of life in the widest sense of the words, which, without him, nature would not have produced. This makes impossible any immediate transference of the laws of life in animal societies to human ones.

The distinction Engels draws between humans and animals is dubious – fungus-growing ants, for example, grow and eat specialised fungi that would not have existed without their activity. But, even if it were valid, why should the difference between collecting and producing be so important as to suspend the laws of evolution? Why would productive capacities not also be susceptible to evolutionary pressures? Engels leaves these questions unanswered.

The dream of perfectibility

Belief in the malleability of human nature has been important for the left because it has provided grounds for hoping that a very different kind of human society is possible. Here, I suspect, is the ultimate reason why the left rejected Darwinian thought. It dashed the left's Great Dream: The Perfectibility of Man. Since Plato's *Republic* at least, the idea of building a perfect society has been present in Western consciousness. For as long as the left has existed, it has sought a society in which all human beings live harmoniously and cooperatively with each other in peace and freedom. Marx and Engels were scornful of 'utopian socialists' and insisted that their own form of socialism was not utopian. They meant by this, however, only that they had discovered the laws of human historical development that would lead to the communist society, and that therefore their socialism was 'scientific', which in their terms meant that it was not utopian. According to these laws of historical development, the class struggle that was driving history would be ended by the victory of the proletariat, and the future communist society would be:

> ... the genuine resolution of the antagonism between man and nature and between man and man; it is the true resolution of the conflict between existence and essence, objectification and self-affirmation, freedom and necessity, individual and species. It is the riddle of history solved ...

This conception of communist society is as firmly utopian as the blueprints for a future society drawn up by Saint-Simon, Fourier or any of the other 'utopian' socialists whose ideas Marx and Engels scorned. Marx wrote that passage as a young man, and some would say that he changed his idea of communism; but, although his terminology became less Hegelian, there is nothing in his later writings to suggest that he abandoned his youthful vision of the future society and much to indicate that he did not. The ethical principle 'from each according to his ability, to each according to his needs' is from one of his last references to communist society and is still firmly in the utopian tradition.

Marx wrote the earlier passage fifteen years before Darwin published *The Origin of Species*, so it was not surprising that he objected to Darwin reading 'the Malthusian "struggle for existence"' into nature. From the start, Marx and Engels recognised the opposition between their own views and the theory of population put forward by Thomas Malthus. The very first work that either of them wrote on economics, Engels' 1844 essay 'Outlines of a Critique of Political Economy', includes a rebuttal of Malthus. In the same year Marx criticised the English Poor Law for seeing pauperism 'as an eternal law of nature, according to the theory of Malthus'. In contrast, Marx and Engels themselves saw poverty as the result of particular economic systems rather than as an inevitable consequence of the workings of nature. Malthus was easy enough to refute, for he offered no good grounds for his assumption that, while population increases geometrically, the food supply can increase only arithmetically. Darwin's theory of evolution, however, was a different matter, and neither Marx nor Engels wished to reject it as a whole. Nevertheless, if the theory of evolution applied to human history in so far as humans are evolved beings, as well as to natural history, the antagonisms and conflicts that Marx saw communism as resolving – the 'antagonism between man and nature, between man and man' and the conflict between 'individual and species' – would never be fully resolved, even though we may learn to make them less destructive. For Darwin the struggle for existence, or at least for the existence of one's offspring, is unending.

This is a long way from the dream of perfecting mankind.

If, on the other hand, the materialist theory of history is correct, and social existence determines consciousness, then the greed, egoism, personal ambition and envy that a Darwinian might see as inevitable aspects of our nature can instead be seen as the consequence of living in a society with private property and private ownership of the means of production. Without these particular social arrangements, people would no longer be so concerned about their private interests. Their nature would change and they would find their happiness in working cooperatively with others for the communal good. That is how communism would overcome the antagonism between man and man. The riddle of history can be solved only if this antagonism is a product of the economic basis of our society, rather than an inherent aspect of our biological nature.

Hence the resolute determination of many on the left to keep Darwinian thinking out of the social arena. Plekhanov, the leading nineteenth-century Russian Marxist, followed Engels in holding that 'Marx's inquiry begins precisely where Darwin's inquiry ends', and this became the conventional wisdom of Marxism. Lenin said that 'the transfer of biological concepts into the field of the social sciences is a meaningless phrase'. As late as the 1960s, school children in the Soviet Union were still taught the simple slogan: 'Darwinism is the science of biological evolution, Marxism of social evolution'. In the same period, the Soviet anthropologist M. F. Nesturkh wrote of the study of human origins that it is the 'sacred duty' of Soviet anthropology 'to consider hominids as people actively forming themselves rather than as animals stubbornly resisting their transformation into human beings'. It is intriguing how two very different ideologies – Christianity and Marxism – agreed with each other in insisting on the gulf between humans and animals, and therefore that evolutionary theory cannot be applied to human beings.

Lysenko, incidentally, went even further in revising Darwinian thinking than those Marxists who denied its application to human affairs. He rejected the idea of competition within species even in nature:

> How explain why bourgeois biology values so highly the 'theory' of intraspecific competition? Because it must justify the fact that, in the capitalist society, the great majority of people, in a period of overproduction of material goods, lives poorly . . . There is no intraspecific competition in nature. There is only competition between species: the wolf eats the hare; the hare does not eat another hare, it eats grass.

Old tunes keep coming back

The conflict between the Marxist theory of history and a biological view of human nature has continued into the late twentieth century. There have been exceptions, most notably J. B. S. Haldane, a major contributor to the development of modern Darwinism and at the same time a communist who did not shy away from acknowledging the influence of evolution and heredity on human affairs. Leading modern evolutionary theorists John Maynard Smith (a student of Haldane) and Robert Trivers have also been involved in left-wing politics.

But consider the following passage:

> . . . determinists assert that the evolution of societies is the result of changes in the frequencies of different sorts of individuals within them. But this confuses cause and effect. Societies evolve because social and economic activity alter the physical and social conditions in which these activities occur. Unique historical events, actions of some individuals, and the altering of consciousness of masses of people interact with the social and economic forces to influence the timing, form, and even the possibility of particular changes; individuals are not totally autonomous units whose individual qualities determine the direction of social evolution. Feudal society did not pass away because some autonomous force increased the frequency of entrepreneurs. On the contrary, the economic activity of Western feudal society itself resulted in a change in economic relations which made serfs into peasants and then into landless industrial workers with all the immense changes in social institutions that were the result.

This clear statement of the materialist theory of history, redolent with the terminology of Marx and Engels, appeared in *BioScience*, the journal of the American Institute of Biological Sciences in March 1976. Its authors were the members of the Sociobiology Study Group of Science for the People, and included the population geneticist Richard Lewontin and other notable figures in the biological sciences. It was written as a response to the emergence of 'sociobiology', which it described as 'another biological determinism'. In its focus on social and economic causes of 'social evolution', it perpetuates the standard Marxist idea of Darwin for natural history and of Marx for human history.

CHAPTER 2

....

Can the Left Accept a Darwinian View of Human Nature?

Unpopular ideas

In the twentieth century the dream of the perfectibility of humankind turned into the nightmares of Stalinist Russia, China under the Cultural Revolution, and Cambodia under Pol Pot. From these nightmares the left awoke in turmoil. There have been attempts to create a new and better society with less terrible results – Castro's Cuba, the Israeli kibbutzim – but none that are unqualified successes. The dream of perfectibility should be put behind us, and, with that, one barrier to a Darwinian left has been removed.

Another barrier that is already partially dismantled is the intellectual hold of Marx's theory of history. Marx has, quite properly, had an immense influence on the social sciences. In directing our attention to the links between a society's economic base and its laws, religion, politics, philosophy and culture generally, Marx shattered the illusion of the independence of ideas and culture, and opened up new and very fruitful areas for research. We should not abandon Marx's insight, but we should make it part of a much larger picture.

Changes in the mode of production in a society do affect its dominant ideas and culture. But to focus only on the differences that these changes can make, and ignore what remains constant, is like looking at the differences in military tactics as weapons have changed over the centuries, without ever asking why nations go to war. It is time to recognise that the way in which the mode of production influences our ideas, our politics, and our consciousness is through the specific features of our biological inheritance.

The remaining great barrier to the acceptance of Darwinian thinking by the left is the idea of the malleability of human nature. Though, as we have seen, this idea is supported by the materialist theory of history, the idea of malleability influences many on the left who are very far from being Marxists, and therefore it can survive the rejection of Marx's theory of history. Non-Marxist reformers often accept a tradition of malleability that stems from John Locke's idea of the mind as 'white Paper, void of all characters, without any Ideas'. From this it follows that education in the broadest possible sense is the great panacea, with the potential to mould human beings into perfect citizens. Beliefs about the malleability or otherwise of human nature tend to vary along the left–right continuum. Yet they are beliefs about a matter of fact and should be open to revision in the light of the evidence.

That evidence can come from history, anthropology, ethology and evolutionary theory. But it is not easy to look at the evidence without ideological blinkers. The Australian anthropologist Derek Freeman discovered this in 1983 when he published *Margaret Mead and Samoa: The Making and Unmaking of an Anthropological Myth*. In *Coming of Age in Samoa*, Mead had argued, on the basis of her observations of Samoan society, 'that adolescence is not necessarily a time of stress and strain, but that cultural conditions make it so . . .'. This conclusion was consistent with the ideas of her teacher, Franz Boas, that our social environment shapes our lives more than any biological factors. Freeman compiled a convincing case that Mead had misunderstood Samoan customs, which did not allow the kind of easy-going adolescent sexuality that Mead described. For this he was pilloried by his fellow-anthropologists, and the American Anthropological Association passed a motion denouncing his refutation of Mead's work as 'unscientific'. Four years later, Freeman's work was further vindicated when one of Mead's original informants came forward and admitted that when Mead questioned her and one of her friends about their sexual behaviour, they had played a prank on her by making up fantastic tales.

The suggestion that there are relatively fixed aspects

of human nature may not be as controversial today as it was twenty-five years ago when E. O. Wilson published *Sociobiology: The New Synthesis*, or even as recently as fifteen years ago, when Freeman's critique of Mead appeared. Since then many books have been published outlining universal aspects of human nature, and they have had far less stormy births. In editing a book on ethics a few years ago, I included a section headed 'Common themes in primate ethics', with readings about altruism towards kin, reciprocity and sexual behaviour in chimpanzees placed alongside readings from anthropologists about similar behaviour in human beings, biblical injunctions against adultery, and passages from Confucius, Hillel and Luther. I expected critics to denounce the idea that nonhuman primates have any such thing as an ethic, let alone that there are common themes in human and nonhuman ethics. But there was no such criticism and the comments I have received on this section have been uniformly favourable. So in the hope that we have matured in our willingness to consider such ideas, I shall now advance the entirely unoriginal proposition that while some areas of human life show great diversity, in others, human behaviour stays fairly constant across the whole range of human cultures, and some aspects of our behaviour are also shared with our closest nonhuman relatives.

What is fixed and what is variable in human nature?

It might help further discussions if we were to try to sort areas of human life into three categories: behaviour that shows great variation across culture, behaviour that shows some variation across culture, and behaviour that shows little or no variation across culture. To do this at anything less than book-length is necessarily a speculative and somewhat arbitrary exercise. Other forms of behaviour could easily have been chosen, and each one could be the subject of an extended debate. Moreover, we should bear in mind that, even where behaviour varies greatly across cultures, this may be the result of fixed psychological rules leading to different outcomes when applied in varied circumstances. Nevertheless, such speculations do have a point: they display the possibilities that anyone interested in Darwinian politics needs to take into account.

In the first category, that of great variation, I would put the way we produce our food – by gathering and hunting, by grazing domesticated animals, or by growing crops. With this would go nomadic or settled lifestyles, and the kinds of food we eat. We could also put economic structures into this category, and religious practices, and forms of government – but not, significantly, the existence of some form of government or group leadership, which seems to be universal, or nearly so.

In the second category, showing some variation, I would put sexual relationships. Victorian anthropologists were most impressed by the differences between attitudes to sexuality in their own society and those in the societies they studied, and as a result we tend to think of sexual morality as an example of something that is totally relative to culture. There are, of course, important differences between societies that allow men to have only one wife, and those that allow men to have more than one wife, but virtually every society has a system of marriage that implies restrictions on sexual intercourse outside the marriage. Moreover, while men may be allowed one wife or more, according to the culture, systems of marriage in which women are allowed to have more than one husband are extremely rare, and usually exist only under special and temporary circumstances. Whatever the rules of marriage may be, and no matter how severe the sanctions, infidelity and sexual jealousy also seem to be universal elements of human sexual behaviour.

Into this category I would also place ethnic identification and its converse, xenophobia and racism. I have the great good fortune of having lived most of my life in a multicultural society with a relatively low level of racism; but I know that racist feelings do exist among a significant number of Australians, and they can be stirred up by demagogues. The tragedy of the Balkans has shown only too vividly how ethnic hatred can be revived among people who have lived peacefully with each other for decades. Racism can be learned and unlearned, but racist demagogues hold their torches over highly flammable material.

In the third category, of little variation across cultures, I would place the fact that we are social beings – not the particular form of society, which may vary greatly, but the fact that human beings, unlike, say, orangutans, generally do not live alone. Equally invariant is our concern for our kin. Our readiness to form cooperative relationships, and to recognise reciprocal obligations, is another universal. More controversially, I would claim that the existence of a hierarchy or system of rank is a near-universal human tendency. There are very few human societies without differences in social status, and when attempts are made to abolish such differences they tend to re-emerge quite rapidly. Finally, sex roles also show relatively little variation. Women almost always have the major role in caring for young children, while men are much more likely than women to be involved in physical conflict, both within the social group and in warfare between groups. Men also tend to have a disproportionate role in the political leadership of the group. So there are common human tendencies that transcend cultural variation.

Of course, culture does have an influence in sharpening or softening even those tendencies that are most deeply rooted in our human nature. There are also variations between individuals. Nothing I have said is contradicted by the existence of individuals who do not care for their kin, or of couples in which the man looks after the children while the woman serves in the army, for the same reason that the statement 'men are generally taller than women' is not contradicted by a tall woman.

I must also stress that my rough classification of human behaviour carries no evaluative overtones. I am not saying that because something like hierarchy, or male dominance, is characteristic of almost all human societies, that therefore it is good, or acceptable, or that we should not attempt to change it. Until quite recent times it has also been characteristic of all human societies for a substantial proportion of women to die in child-birth, but no one doubts that our success in changing this has been a good thing. My point is not about deducing an 'ought' from an 'is' but about gaining a better understanding of what it may take to achieve the goals we seek.

To be blind to the facts about human nature is to risk disaster. Consider hierarchy. To say that human beings under a wide range of conditions have a tendency to form hierarchies is not to say that it is right for our society to remain hierarchical; but it is to issue a warning that we should not expect to abolish hierarchy by eliminating the particular hierarchy we have in our society. For example, if we live in a society with a hierarchy based on a hereditary aristocracy, and we abolish the hereditary aristocracy, as the French and American revolutionaries did, we are likely to find that a new hierarchy emerges based on something else, perhaps military power or wealth. When the Bolshevik Revolution in Russia abolished both the hereditary aristocracy and private wealth, a hierarchy soon developed on the basis of rank and influence within the Communist Party, and this became the basis for a variety of privileges. The tendency to form hierarchies shows itself in all sorts of petty ways in corporations and bureaucracies, where people place enormous importance on how big their office is, and how many windows it has. Position in a hierarchy even appears to have an impact, independently of other variables, on health and longevity.

Seeing this tendency to form hierarchies as inherent in human beings helps us to understand the rapid departure from equality in the Soviet Union. It is easy to say, as Trotsky did, that Stalin 'betrayed' the revolution. But before Stalin came to power, Trotsky himself, with Lenin's full support, had surely betrayed the idea of communism as a liberating force in his brutal suppression of the Kronstadt rebellion. In fact the Communist movement had been dominated by authoritarian figures ever since Marx himself, who destroyed the First International when he saw that his opponents were likely to win control over it. But the point is not what individual revolutionaries do. The point is: what egalitarian revolution has not been betrayed by its leaders? And why do we dream that the next revolution will be any different? None of this shows that hierarchy is good, or desirable, or even inevitable, but it does show that getting rid of it is not going to be nearly as easy as revolutionaries usually imagine. These are the facts that the left needs to grapple with. To do so, the left has to accept and understand our nature as evolved beings.

How can reformers learn from Darwin?

Wood carvers presented with a piece of timber and a request to make wooden bowls from it do not simply begin carving according to a design drawn up before they have seen the wood. Instead they will examine the material with which they are to work, and modify their design in order to suit its grain. Political philosophers and the revolutionaries or reformers who have followed them have all too often worked out their ideal society, or their reforms, and sought to apply them without knowing much about the human beings who must carry out, and live with, their plans. Then, when the plans don't work, they blame traitors within their ranks, or sinister agents of outside forces, for the failure. Instead, those seeking to reshape society must understand the tendencies inherent in human beings, and modify their abstract ideals in order to suit them.

There are different ways of working with the tendencies inherent in human nature. The market economy is based on the idea that human beings can be relied upon to work hard and show initiative only if by doing so they will further their own economic interests. As Adam Smith put it, 'It is not from the benevolence of the butcher that we expect our dinner, but from his regard to his own self-interest.' To serve our own interests we will strive to produce better goods than our competitors, or to produce similar goods more cheaply. Thus, Smith said, the self-interested desires of a multitude of individuals are drawn together, as if by a hidden hand, to work for the benefit of all. Garrett Hardin put this view succinctly in his book *The Limits of Altruism* when he wrote that public policies should be based on 'An unwavering adherence to the Cardinal Rule: Never ask a person to act against his own self-interest.' The current economic fashion for privatising enterprises and introducing competition into areas that were previously state monopolies is consonant with this way of thinking. In theory – abstract theory that is, without any assumptions about human nature – a state monopoly should be able to provide the cheapest and most efficient utility services or transport services or, for that matter, bread supply, since such a monopoly would have enormous advantages of scale, and would not have to make profits for its owners. When we factor in the popular assumption that self-interest – or more specifically, the desire to enrich oneself – is what drives human beings to work well, however, the picture changes. If the community owns an enterprise, its managers do not profit from its success. Their own economic interest and that of the enterprise don't necessarily coincide, and the result can be, at best, inefficiency and, at worst, widespread corruption and theft. Privatising the enterprise will mean that the owners ensure that management is rewarded in accordance with performance, and so the managers will take steps to ensure that the enterprise runs as efficiently as possible.

That is one way of tailoring our institutions to human nature, or at least to one view of human nature. But it is not the only way of doing so. Even within the terms

41

of Hardin's Cardinal Rule, we still have to ask what we mean by the term 'self-interest'. We often assume that it is in our interests to earn as much money as possible, but there is no reason to assume that earning more than a modest amount of money will maximise the number of descendants we leave in future generations. So from an evolutionary perspective, we cannot identify self-interest with wealth. Nor can we do this from a more commonsense viewpoint. We often hear it said that money cannot buy happiness. This may be trite, but it carries the implication that it is more in our interests to be happy than to be rich. Properly understood, self-interest is broader than economic self-interest. Most people want their lives to be happy, fulfilling, or meaningful in some way, and they recognise that money is, at best, a means to achieving part of these ends. Public policy does not have to rely on self-interest in this narrow economic sense. It can, instead, appeal to the widespread need to feel wanted, or useful, or to belong to a community – all things that are more likely to come from cooperating with others than from competing with them.

Modern Darwinian thought embraces both competition and reciprocal altruism, which is really a more technical term for cooperation. Focusing largely on the competitive element, modern market economies are premised on the idea that we are all dominated by acquisitive and competitive desires. These economies are continuing to design their structures to channel our acquisitive and competitive desires so that they work for the good of all. That is, undoubtedly, better than a situation in which they work only for the good of a few. But even when the competitive consumer society works at its best, it is not the only way of harmonising our nature and the common good. Instead we could seek to encourage a broader sense of our interests, in which we seek to build on the social and cooperative side of our nature, in addition to the individualistic and competitive side.

CHAPTER 3

....

Competition or Cooperation?

Building a more cooperative society

Any human society will show some competitive tendencies and some cooperative ones. We cannot change
that, but we may be able to change the balance between
these two elements. America in the twentieth century
has been the paradigm of a competitive society, in which
the drive for personal wealth and to get to the top is
widely seen as the goal of everything we do. In Japan,
people tend to be more group-oriented and much less
likely to push themselves forward. The popular Japanese
saying 'The nail that sticks out gets hammered down' is
the antithesis of the Western proverb 'It's the squeaky
wheel that gets the oil'. Yet in America it is easy to find

examples of cooperation, and in Japan people know how to assert themselves and compete with rivals in a manner that is socially acceptable. So in discussing societies that encourage competition rather than cooperation, and vice versa, I am talking about societies that differ in degree, rather than in kind; and the form of competition I have in mind is primarily competition in the marketplace, and hence is ultimately competition for personal wealth.

A cooperative society is more in keeping with the values of the left than a competitive society. Encouraging the pursuit of self-interest through the free market may have contributed to a high average level of prosperity in the developed countries, but at the same time the gap between rich and poor has widened, and support for the poor has been cut back. Any principle of justice that gives precedence to raising the level of the worst-off groups in society would have to regard this as indefensible; and from a utilitarian perspective I doubt that greater wealth for the middle and upper classes could make up for the increase in human misery that this has brought to the poor. Both national and international surveys show little correlation between an increase in wealth and an increase in happiness, once basic needs have been met. The competitive society, carried to extremes, encourages us to see self-interest in the terms of a T-shirt once owned by the Wall Street entrepreneur Ivan Boesky (the putative model for Gordon Gekko in Oliver Stone's film *Wall Street*): 'He who has the most toys when he dies, wins.' But it takes only a little reflection to realise that a society in which people are primarily motivated by a desire to keep up with, or ahead of, their neighbours is not likely to be a society in which most citizens find happiness and fulfilment. There are other costs, too, in living in a society divided between rich and poor, even if you are lucky enough to be among the rich. As Robert Bellah and his colleagues put it in *Habits of the Heart*: 'One cannot live a rich private life in a state of siege, mistrusting all strangers and turning one's home into an armed camp.'

How can we build a society that is cooperative and offers a strong safety net for those who are unable to provide for their own needs? As I have already mentioned, a readiness to cooperate seems to be part of our nature. Humans have shown that they are able to see the benefits of cooperation in what would appear to be the most unpromising circumstances. In the stalemate of trench warfare in Northern France in the first year of World War I, between the pitched battles in which tens of thousands died, ordinary soldiers of the opposing armies developed the extraordinary system known as 'live and let live'. In tacit defiance of the orders of their superiors, the troops facing each other across no man's land tried, and usually succeeded, in not killing each other. Tests of our ability to make inferences show that although we are not adept at formal logic, we are particularly good at recognising social contracts, and especially, the cheats who break them. This readiness to cooperate is a true universal among humans (and not only among humans — it applies to other long-lived intelligent social animals too). There is now a huge literature on this topic, and I will not attempt to summarise it here. But I shall consider Robert Axelrod's work in game theory, not only because it is entirely compatible with a Darwinian perspective (indeed Axelrod's book is called *The Evolution of Cooperation* and one chapter is co-authored by the leading evolutionary theorist W. D. Hamilton), but also because Axelrod's pioneering work explores the circumstances in which cooperation can flourish. It should be seized upon as a starting point for the development of a field of social research that shows the way towards a more cooperative society.

The prisoner's dilemma

Axelrod's work starts off as an investigation of a well-known puzzle about cooperation called the prisoner's dilemma. There are many versions of the dilemma. Here is mine.

You and another prisoner are languishing in separate cells of the Ruritanian Police Headquarters. You have no way of contacting him. The police are trying to get you both to confess to plotting against the state. An interrogator offers you a deal: if the other prisoner continues to remain silent, but you confess, implicating him in your crime, you will be freed and he will be locked away for twenty years. If, however, you refuse to confess, and the other prisoner does, you will be the one who gets twenty years and he will go free. You ask:

'What if we both confess?' The interrogator says that then you will both get ten years. 'And if neither of us confesses?' Reluctantly, the interrogator admits that he will not be able to get convictions, but they can, and will, hold you for another six months under the government's emergency powers legislation. 'But think about it,' he adds, 'whether the other guy sings or not, you'll be better off if you do – you can walk straight out of here if he doesn't, which is better than staying here for another six months, and you get ten years rather than twenty if he does. And remember, we've offered him exactly the same deal. So what are you going to do?'

The prisoner's dilemma is whether to confess or not. Assuming that the prisoner wishes to spend the minimum time in prison, it would seem to be rational to confess. No matter what the other prisoner does, that will be to the advantage of the confessing prisoner. But each prisoner faces the same dilemma, and if they both follow their own individual interests and confess, they will end up serving ten years, when they could have both been out in six months!

There is no solution to this dilemma. It shows that the outcome of rational, self-interested choices by two or more individuals can make all of them worse off than they would have been if they had not pursued their own short-term self-interest. The individual pursuit of self-interest can be collectively self-defeating.

Really, we shouldn't be surprised by such a result. People who commute to work by car face it every morning. They would all be better off if, instead of sitting in heavy traffic, they abandoned their cars and used the buses, which would then travel swiftly down uncrowded roads. But it is not in the interests of any individual to switch to the bus because, as long as most people continue to use their cars, the buses will be even slower than the cars. Arms races also have the same logic as the prisoner's dilemma: it is not in the interests of either side to disarm while the other does not, but both would be better off if they did not have to spend so much on weapons. There are many other examples in real life. In most of them, it is possible to change the pay-offs, or coordinate the behaviour of those involved, to bring about a better outcome. Traffic planners can put in bus lanes; arms negotiators can have inspection teams. The prisoner's dilemma is a pure example of the problem, because the pay-offs cannot be changed, the prisoners cannot coordinate their behaviour, and it is a once-in-a-lifetime situation. Under such conditions the cooperative strategy is not justified in terms of self-interest. Paradoxically, only altruism can help the prisoners. They will each be better off if each of them takes into account, not only how long he will spend in prison, but how long the other prisoner will be there. Then they can each see that to minimise the total amount of time spent in prison by the two of them, neither should confess.

Although a once-in-a-lifetime prisoner's dilemma is insoluble, the situation changes when the same parties find themselves faced with repeated prisoner's dilemma-type choices. Then, even though in each choice the participants can do better by not cooperating, over time cooperation may be the better strategy. To test this idea, and to discover exactly what strategy would have the best pay-offs, Axelrod invited people interested in game theory to send in their suggestions for what strategy would produce the best pay-off for the person using it, if they were in repeated prisoner's dilemma situations. If we use the term 'cooperate' to describe remaining silent in the classic prisoner's dilemma situation, and the term 'defect' as a label for confessing, then the possible strategies range from 'Always cooperate' to 'Always defect', with an infinite number of possibilities in between, including random choice, and choices that respond in some way to what the other prisoner has done previously. When Axelrod had received the strategies, he ran them against each other on a computer, in a kind of round-robin tournament in which each strategy was pitted against each other strategy 200 times. The computer recorded the outcomes, which of course depended on what the other strategy did each time. The winner was a simple strategy called 'Tit for Tat'. It opened every encounter with a new 'prisoner' by cooperating. After that, it simply did whatever the prisoner had done the previous time. So if the other prisoner cooperated, it cooperated, and it continued to cooperate unless the other prisoner defected. Then it defected too, and continued to do so unless the other prisoner again cooperated. Tit for Tat also won a second tournament that Axelrod organised, even though the people sending in strategies this time knew that it had won the previous tournament, and were trying to beat it.

Learning from Tit for Tat

The success of a strategy in a computer simulation tournament may seem a long way from real life, but the left can learn from Axelrod's work about how to build a more cooperative society.

Axelrod's results, which have generally been supported by subsequent work in the field, can serve as a basis for social planning that should appeal to the left. His findings suggest that we should be able to create conditions that draw on our inherent grasp of the rules of mutually beneficial cooperation and thereby make it possible for mutually beneficial relationships to flourish where otherwise they would not. Here we have an example of the interaction between our biology and our social environment that shows clearly that Darwinian ways of thinking do not assume short-term selfishness.

Members of the more idealistic left may regret that those following Tit for Tat do not continue to cooperate no matter what. A left that understands Darwin will know that this is because niches tend to be filled. If there are grasses to be eaten, herbivores will evolve capable of eating them. If there are herbivores to be eaten, predators will evolve to prey on them. In human society, if there are ways of making an easy living, there will be people who will find them. In Richard Dawkins' terminology, if there are 'suckers' there will also be 'cheats' who can prosper by taking advantage of them. The 'sucker' need not be an individual – it could be an institution or even the state. The easier it is for cheats to make a living, the more of them there are likely to be. A pre-Darwinian left would blame the existence of cheats on poverty, or a lack of education, or the legacy of reactionary capitalist ways of thinking. A Darwinian left will realise that while all of these factors may make a difference to the level of cheating, the only permanent solution is to change the pay-offs so that cheats do not prosper. This means not turning the other cheek.

We need to think about how to set up the conditions in which cooperation thrives. The first problem to deal with is that of scale. Tit for Tat cannot work in a society of strangers who will never encounter each other again. No wonder that people living in big cities do not always show the consideration to each other that is the norm in a rural village in which people have known each other all their lives. What structures can overcome the anonymity of the huge, highly mobile societies that have come into existence in this century and show every sign of increasing in size with the globalisation of the world economy?

The next problem is, if anything, even more difficult to overcome. If nothing that you do really makes much difference to me, Tit for Tat will not work. So while equality is not required, too great a disparity in power or wealth will remove the incentive for mutual cooperation. This strongly suggests a need to do something about the economic trends in developed nations which for the past decade or more have increased economic inequality. The left, of course, has ample reason to reverse these trends and to make people at the bottom better off. But the aim of creating a society based on mutually beneficial cooperation adds another strong reason to do so: to leave a group of people so far outside the social commonwealth that they have nothing to contribute to it, is to alienate them from social practices and institutions in a manner that almost ensures that they will become adversaries who pose a danger to those institutions. The political lesson of twentieth-century Darwinian thinking is, therefore, entirely different from that of nineteenth-century Social Darwinism. Social Darwinists saw the fact that those who are less fit will fall by the wayside as nature's way of weeding out the unfit, and an inevitable result of the struggle for existence. To try to overcome it or even ameliorate it was futile, if not positively harmful. A Darwinian left, understanding the prerequisites for mutual cooperation as well as its benefits, would strive to avoid economic conditions that create outcasts. At a time when some sections of the left are accepting a narrowly economic view of social policy, an evolutionary view of human psychology can show us the potential social cost of writing off those whose labour might, in purely economic terms, not be worth employing. When the free operation of competitive market forces makes it hazardous to walk the streets at night, governments do well to interfere with those market forces to promote employment. This may be done by means of tax concessions, subsidies or direct employment for socially useful work, depending on what will work better to bring those on the margins of society back into the mainstream.

CHAPTER 4

....

From Cooperation to Altruism?

The puzzle of the evolution of altruism

A society that fosters cooperation can take the left some distance towards its goals. But would a Darwinian left really have to accept Hardin's Cardinal Rule that we should never ask people to act against their own self-interest? Is it really impossible for the left to seek to promote a society in which there is a strong feeling of concern for the good of others, and opportunities for people to work for the good of the whole society or, even more broadly, for the welfare of human and nonhuman sentient beings wherever they may live? Altruism – not just kin altruism, or reciprocal altruism, but genuine altruism toward strangers – does exist.

Nearly thirty years ago, in *The Gift Relationship*, Richard Titmuss pointed out that the Blood Bank survives on the altruism of its donors, and provides a means by which a person can give something precious to a stranger without making it possible for the stranger to return the favour in any way. (In Britain, as in many other countries with voluntary blood donors and national provision of health care, the provision of blood to a patient who needs it is not linked in any way to whether the patient has ever donated blood, so this is a genuinely altruistic – and often private – act that gains no reward.) Voluntary blood banks have come under strain as the demand for blood increases, and some donors have to be ruled ineligible because of the risk of spreading disease. But, on the whole, the system has survived. This altruistic practice takes place in a society that does little to encourage altruistic tendencies and, in some respects, by its promotion of individualist competitiveness, positively discourages them.

Darwinian thinking suggests that we are not likely to be naturally altruistic. How can a self-sacrificing trait that benefits the group at the expense of the individual survive? It seems that self-sacrificing tendencies would be eliminated from the gene pool, no matter how much they help the group. But there may be selective forces that encourage behaviour that looks like altruism, and may be altruistic in its motivation, even though in specific circumstances it brings benefits to the apparently altruistic individual. A variety of social practices, from peer group esteem to government policies, can play an important role here, by providing rewards and punishments for behaviour that benefits or harms others. Sometimes the rewards are quite transparently a way of making up for what would otherwise be a loss of reproductive success. Among some Native American tribes of the Great Plains, a warrior might make a solemn vow to fight to the death in a forthcoming battle. He would then be allowed, in the days leading up to the fight, to make love to as many willing women as he wished. When, during the World War I, girls handed white feathers to men of military age in civilian clothes, they were surely making the same kind of point, even if the reproductive pay-off was not quite so direct. Conversely, when behaviour that harms others is punished with life imprisonment or death, the drastic negative impact on reproductive success is clear. But, more commonly, behaviour that is perceived as good will be rewarded in terms of enhanced popularity and social standing, while behaviour that is perceived as bad will be punished by disapproval, reduced popularity, and perhaps ostracism. (These pressures to conform to group standards of what constitutes good or bad behaviour can, of course, be used for evil purposes as well as for good ones.)

Granted, to the extent that those who benefit others gain a reward or avoid a punishment, they are not acting 'altruistically', in the sense used in evolutionary theory. It is also true that, to the extent that they are motivated by the prospect of obtaining a reward or avoiding a punishment, they are not acting altruistically in the normal sense of that term, which focuses on the motivation of an action rather than its impact on reproductive fitness. But the two senses are distinct, and it is possible for an action to be altruistic in the everyday sense of the term, but not in the sense used in evolutionary theory. A person who volunteers to work with intellectually disabled children *may* thereby become more attractive to the opposite sex and so increase his or her reproductive fitness, but if the person volunteered only in order to help the children have a better life, without any thought or expectation of attracting the opposite sex, then the volunteer is acting altruistically in the everyday sense, whatever the consequences of the action may be. The everyday sense of altruism deals with conscious motives and expectations. We are quite good at distinguishing those who genuinely want to help others from those who do it for ulterior motives. We often reserve the greatest rewards for those who do not seek them, precisely because we wish to encourage readiness to sacrifice one's own interests for the sake of others, when this will clearly be of great benefit from the perspective of the larger group.

Thus it is a mistake to say that evolutionary theory shows that people cannot be motivated by a desire to help others. It shows no such thing, which is just as well for the theory because, as we have seen, altruism towards strangers does exist, in institutions like the Blood Bank. It has to be admitted, though, that the sacrifice asked of a donor is not burdensome, and nevertheless only a small minority of people donate blood – in Britain, around 6 per cent. Bone marrow donation demands more of the donor – general anaesthesia, a night in hospital, and some discomfort for a time afterwards. The proportion of people registering their willingness to donate bone marrow to a stranger is, not surprisingly, lower than the number who will give blood; but in England it still amounts to more than 180,000 people. On the evidence we have so far, institutions based on altruism to strangers can work, as long as they do not rely on a large proportion of the population to act altruistically – and the greater the sacrifice asked, the smaller the proportion of the population that we should expect to respond to it. But we need to understand more about what it is that leads people to donate blood or bone marrow, so that we can base social policies on a more secure foundation of knowledge about human behaviour.

Status for what?

We live in a competitive society that values consumption and relates status to media interest. In such a society there is little connection between status and the benefits one brings to others. In *The Theory of the Leisure Class*, written a century ago, Thorstein Veblen showed how the wealthy use 'conspicuous consumption' and 'conspicuous waste' to display their status. Tom Wolfe's *Bonfire of the Vanities* shows nicely how little has changed. In one brilliant vignette, he describes how the bond trader Sherman McCoy and his wife Judy travel to a dinner party only six New York blocks from where they live. Unfortunately Judy's dress is not designed for walking in the street. Veblen would have understood that well, because, as he put it:

> Our dress . . . in order to serve its purpose effectually, should not only be expensive, but it should also make plain to all observers that the wearer is not engaged in any kind of productive labor . . . It may broadly be set down that the womanliness of woman's apparel resolves itself, in point of substantial fact, into the more effective hindrance to useful exertion offered by the garments peculiar to women.

So walking is out. Nor is getting a taxi possible, because it would be socially unbecoming for the other guests to see the McCoys out on Fifth Avenue, trying to hail a taxi to get home. So they must hire a limousine and driver to take them six blocks, wait four hours, and drive them home again.

Can we strengthen concern for others by shifting ideas of status away from conspicuous consumption, in a more socially desirable direction? With all the economic forces favouring the promotion of high levels of consumption, it wouldn't be easy, but there are several thinkers working in this direction. In *The Winner-Take-All-Society*, Robert Frank and Phillip Cook argue that a tax on spending – payable through our tax returns rather than as part of the price we pay when we buy – would have a significant positive impact in changing the habits of the high-fliers. Whether or not they are right, this idea, and others in the same arena, are ripe for further investigation.

CHAPTER 5

....

A Darwinian Left for Today and Beyond

This short book has been a sketch of the ways in which a Darwinian left would differ from the traditional left that we have come to know over the past two hundred years. In closing, I shall first draw together, in point form, some of the features that I think would distinguish a Darwinian left from previous versions of the left, both old and new; these are features that I think a Darwinian left should embrace today. Then I will cast a glance at more distant prospects.

A Darwinian left would not:

• Deny the existence of a human nature, nor insist that human nature is inherently good, nor that it is infinitely malleable;

- Expect to end all conflict and strife between human beings, whether by political revolution, social change, or better education;

- Assume that all inequalities are due to discrimination, prejudice, oppression or social conditioning. Some will be, but this cannot be assumed in every case;

A Darwinian left would:

- Accept that there is such a thing as human nature, and seek to find out more about it, so that policies can be grounded on the best available evidence of what human beings are like;

- Reject any inference from what is 'natural' to what is 'right';

- Expect that, under different social and economic systems, many people will act competitively in order to enhance their own status, gain a position of power, and/or advance their interests and those of their kin;

- Expect that, regardless of the social and economic system in which they live, most people will respond positively to genuine opportunities to enter into mutually beneficial forms of cooperation;

- Promote structures that foster cooperation rather than competition, and attempt to channel competition into socially desirable ends;

- Recognise that the way in which we exploit nonhuman animals is a legacy of a pre-Darwinian past that exaggerated the gulf between humans and other animals, and therefore work towards a higher moral status for nonhuman animals, and a less anthropocentric view of our dominance over nature;

- Stand by the traditional values of the left by being on the side of the weak, poor and oppressed, but think very carefully about what social and economic changes will really work to benefit them.

In some ways, this is a sharply deflated vision of the left, its utopian ideas replaced by a coolly realistic view of what can be achieved. That is, I think, the best we can do today – and it is still a much more positive view than that which many on the left have assumed to be implied in a Darwinian understanding of human nature.

If we take a much longer-term perspective, there may be a prospect for restoring more far-reaching ambitions of change. We do not know to what extent our capacity to reason can, in the long run, take us beyond the conventional Darwinian constraints on the degree of altruism that a society may be able to foster. We are reasoning beings. In other works I have likened reason to an escalator, in that, once we start reasoning, we may be compelled to follow a chain of argument to a conclusion that we did not anticipate when we began. Reason provides us with the capacity to recognise that each of us is simply one being among others, all of whom have wants and needs that matter to them, as our needs and wants matter to us. Can that insight ever overcome the pull of other elements in our evolved nature that act against the idea of an impartial concern for all of our fellow humans, or, better still, for all sentient beings?

No less a champion of Darwinian thought than Richard Dawkins holds out the prospect of 'deliberately cultivating and nurturing pure, disinterested altruism – something that has no place in nature, something that has never existed before in the whole history of the world'. Although 'We are built as gene machines,' he tells us, 'we have the power to turn against our creators'. There is an important truth here. We are the first generation to understand not only that we have evolved, but also the mechanisms by which we have evolved and how this evolutionary heritage influences our behaviour. In his philosophical epic, *The Phenomenology of Mind*, Hegel portrayed the culmination of history as a state of Absolute Knowledge, in which Mind knows itself for what it is, and hence achieves its own freedom. We don't have to buy Hegel's metaphysics to see that something similar really has happened in the last fifty years. For the first time since life emerged from the primeval soup, there are beings who understand how they have come to be what they are. To those who fear adding to the power of government and the scientific establishment, this seems more of a danger than a source of freedom. In a more distant future that we can still barely glimpse, it may turn out to be the prerequisite for a new kind of freedom.

NOTES AND REFERENCES

....

p. 3 Marx on Bakunin's *Statism and Anarchy*: In *Karl Marx: Selected Writings* (ed. McLellan, D.) p. 563 (Oxford University Press, Oxford, 1977).

p. 9 Combined net worth of richest 400 people: Barnet, R. J. & Cavanagh, J. *Global Dreams: Imperial Corporations and the New World Order* (Simon & Schuster, New York, 1994). Number of people living on less than US$1 per day: World Bank Development Indicators, 1997.

p. 10 Darwin to Lyell: In *The Life and Letters of Charles Darwin* (ed. Darwin. F.) vol. ii, p. 62 (Murray, London, 1887).

p. 11 Carnegie on competition: Carnegie, A. 'Wealth', *North American Review* **391**, 654–7 (1889). Rockefeller on competition: Quoted in Huber, R. *The American Idea of Success*, p. 66 (McGraw-Hill, New York, 1971). Holmes on Fourteenth Amendment: *Lochner* v. *New York* [1905] 198 US 45, quoted in Hofstadter, R. *Social Darwinism in American Thought*, p. 47 (Beacon, Boston, 1966).

p. 12 Hume: Hume, D. *A Treatise on Human Nature*, bk

iii, pt i, sect. i (1739). Wilson: Wilson, E. O. *On Human Nature*, pp. 5, 198–9 (Harvard University Press, Cambridge, Mass., 1978). Einstein: Einstein, A. *Out of My Later Years*, p. 114 (Philosophical Library, New York, 1950).

p. 13 Wilson on Rawls: Wilson, E. O. *Sociobiology: The New Synthesis*, p. 562 (Harvard University Press, Cambridge, Mass., 1975).

p. 14 Wilson on sex differences: Wilson, E. O. *On Human Nature* p. 134 (Harvard University Press, Cambridge, Mass., 1978).

p. 17 Attitudes to animals and relationship to humans: Rachels, J. *Created from Animals: The Moral Implications of Darwinism* (Oxford University Press, Oxford, 1990). Why Darwinian political thinkers should recognise similarities: Singer, P. *Animal Liberation*, 2nd edn (Pimlico, London, 1995).

p. 18 Men and women in the workplace: Browne, K. *Divided Labours: An Evolutionary View of Women at Work* (Weidenfeld & Nicholson, 1998).

p. 19 Tennyson on nature: Tennyson, A. *In Memoriam* (1850). Maynard Smith: Foreword to Cronin, H. *The Ant and the Peacock*, p. xx (Cambridge University Press, Cambridge, 1991). *Mutual Aid*: Kropotkin, P. *Mutual Aid: A Factor of Evolution* (Heinemann, London, 1902).

p. 20 Errors of respected evolutionists: Cronin, H. *The Ant and the Peacock*, pp. 280–1 (Cambridge University Press, Cambridge, 1991). Marx to Lassalle (16 January 1862): In *Karl Marx: Selected Writings* (ed.

McLellan, D.) p. 525 (Oxford University Press, Oxford, 1977).

p. 21 Marx on Darwin's work as product of bourgeois society: Marx to Engels (18 June 1862), in *Karl Marx: Selected Writings* (ed. McLellan, D.) p. 526 (Oxford University Press, Oxford, 1977). 'The Part Played by Labour in the Transition from Ape to Man': In *Karl Marx and Friedrich Engels, Selected Works*, vol. ii, pp. 80–92 (Foreign Languages Publishing House, Moscow, 1962).

p. 22 Lysenko's influence on Soviet agriculture: Medvedev, Z. *The Rise and Fall of T. D. Lysenko*, pp. 7–8, 151–94 (Columbia University Press, New York, 1969). Marx's materialist view of history: Preface to *A Critique of Political Economy*, in *Karl Marx: Selected Writings* (ed. McLellan, D.) p. 389 (Oxford University Press, Oxford, 1977).

p. 23 *Discourse on the Origin of Inequality*: Rousseau, J.-J. *Discourse on the Origin of Inequality*, p. 192 (Dent, London, 1958). *Dialectics of Human Nature*: In *Marx and Engels on Malthus* (ed. Meek, R.) p. 187 (Lawrence & Wishart, London, 1953).

p. 24 Perfectibility of Man: Passmore, J. *The Perfectibility of Man*, 2nd edn (Scribner's, New York, 1970). Engels on 'utopian socialists': Engels, F. *Socialism, Utopian and Scientific*, in *Karl Marx and Friedrich Engels, Selected Works*, vol. ii (Foreign Languages Publishing House, Moscow, 1962).

p. 25 '. . . the genuine resolution': 'Economic and Philosophic Manuscripts of 1844', in *Karl Marx: Selected*

Writings (ed. McLellan, D.) p. 89 (Oxford University Press, Oxford, 1977). Marx on communist society: 'Critique of the Gotha Programme', in *Karl Marx: Selected Writings* (ed. McLellan, D.) p. 566 (Oxford University Press, Oxford, 1977).

p. 26 Marx on the Poor Law: 'Critical Remarks on the Article "The King of Prussia and Social Reform"', in *Marx and Engels on Malthus* (ed. Meek, R.) p. 67 (Lawrence & Wishart, London, 1953).

p. 27 Determination of left to exclude Darwinian thinking from social arena: This paragraph draws on Joravsky, D. *The Lysenko Affair*, pp. 230, 254–58 (Harvard University Press, Cambridge, Mass., 1970). Plekhanov is quoted on p. 258; the original source is Plekhanov, G. V. *Izbrannye filosfskie proizvedeniia*, vol. i, pp. 690–1 (5 vols, Moscow, 1956–58). Lenin is quoted on p. 257; the original source is Lenin, V. I. *Sochineniia*, vol. xiv, p. 315 (4th edn, 44 vols, Moscow, 1941–67). Nesturkh is quoted on p. 270; the original is Nesturkh, M. F. 'Darvin i sovremennye problemy antropogeneza', *Voprosy antropologii*, no. 2, p. 14 (1960).

p. 28 Lysenko on competition: Lysenko, T. D. *Agrobiologiya*, pp. 602–6 (Sel'khozgiz, 1949); quoted in Medvedev, Z. *The Rise and Fall of T.D. Lysenko*, p. 107 (Columbia University Press, New York, 1969). Haldane on evolution and human affairs: Haldane, J. B. S. *Heredity and Politics* (Allen & Unwin, London, 1938).

p. 29 Modern evolutionists in left-wing politics: Segerstrale, U. 'Colleagues in Conflict: An "In Vivo"

Analysis of the Sociobiology Controversy', *Biology and Philosophy* **1:1**, 53–87 (1985); van den Berghe, P. 'Sociobiology: Several Views', *BioScience* **31**, 406 (1980). Statement ot the materialist view of history: Sociobiology Study Group of Science for the People, 'Sociobiology: Another Biological Determinism', *BioScience* **26:3** (March 1976), quoted in Caplan, A. (ed.) *The Sociobiology Debate*, p. 289 (Harper & Row, New York, 1978).

p. 32 Locke on the mind: Locke, J. *An Essay Concerning Human Understanding*, bk 2, ch. 1, para. 2 (1689).

p. 33 *Margaret Mead and Samoa*: Freeman, D. *Margaret Mead and Samoa: The Making and Unmaking of an Anthropological Myth* (Harvard University Press, Cambridge, Mass., 1983). *Coming of Age in Samoa*: Mead, M. *Coming of Age in Samoa*, p. 186 (Penguin, London, 1966). Freeman further vindicated: Freeman, D. *Paradigms in Collision* (Research School of Pacific Studies, Australian National University, Canberra, 1992).

p. 35 Fixed or variable? Documentation on the universality or otherwise of traits mentioned in this section is extensive. A small selection includes: Westermarck, E. *The Origin and Development of the Moral Ideas* (Macmillan, London, 1906); Wilson, E. O. *Sociobiology: The New Synthesis* (Harvard University Press, Cambridge, Mass., 1975); Wilson, E. O. *On Human Nature* (Harvard University Press, Cambridge, Mass., 1978); van den Berge, P. L. *Human Family Systems* (Elsevier, New York, 1979); Alexander, R. D. *Darwinism and Human Affairs* (University of Washington

Press, Seattle, 1979); Symons, D. *The Evolution of Human Sexuality* (Oxford University Press, Oxford, 1979); Cronin, H. *The Ant and the Peacock* (Cambridge University Press, Cambridge, 1991); Wilson, J. Q. *The Moral Sense* (Free Press, New York, 1993); Brown, D. E. *Human Universals* (McGraw-Hill, New York, 1991); Pinker, S. *The Language Instinct* (Morrow, New York, 1994); Pinker, S. *How the Mind Works* (Norton, New York, 1998).

p. 39 Hierarchy, health and longevity: Marmot, M. G. *et al.* 'Health Inequalities Among British Civil Servants: The Whitehall II Study', *Lancet* **I,** 1387–93 (1991); Marmot, M. G. & Theorell, T. 'Social Class and Cardiovascular Disease: The Contribution of Work', *International Journal of Health Services* **18**, 659–74 (1998); Wilkinson, R. *Unhealthy Societies: The Afflictions of Inequality* (Routledge, London, 1996).

p. 40 Smith: Smith, A. *The Wealth of Nations* (ed. Campbell, R. H. & Skinner, A. S.) p. 24 (Clarendon Press, Oxford, 1976).

p. 41 *The Limits of Altruism*: Hardin, G. *The Limits of Altruism: An Ecologist's View of Survival* (Indiana University Press, Bloomington, 1977).

p. 45 Survey of research on wealth and happiness: Singer, P. *How Are We To Live?*, pp. 59–62 (Oxford University Press, Oxford, 1997).

p. 46 *Habits of the Heart*: Bellah, R. *et al. Habits of the Heart*, p. 163 (University of California Press, Berkeley, 1985). The 'live and let live' system: Ashworth, T. *Trench Warfare, 1914–18: The Live and Let Live System*

(Holmes & Meyer, New York, 1980). Tests of our ability to make inferences: Cosmides, L. 'The Logic of Social Exchange', *Cognition* **31**, 187–276 (1989).

p. 47 *The Evolution of Cooperation*: Axelrod, R. *The Evolution of Cooperation* (BasicBooks, 1984).

p. 55 *The Gift Relationship*: Titmuss, R. M. *The Gift Relationship: From Human Blood to Social Policy* (Allen & Unwin, 1971).

p. 58 *The Theory of the Leisure Class*: Veblen, T. *The Theory of the Leisure Class* (Penguin, New York, 1889/1994).

p. 59 *The Winner-Takes-All Society*: Frank, R. & Cook, P. *The Winner-Take-All Society* (Free Press, New York, 1996).

p. 63 Richard Dawkins: *The Selfish Gene*, p. 215 (Oxford University Press, Oxford, 1976).